AN ENGINEERING VIEW

of the

UNIVERSE VOL V – QUESTIONS

by Robert M. Heilman

The Series

AN ENGINEERING VIEW OF THE

UNIVERSE

Books from the point of view of Engineering Methodology.

VOL I – So many things wrong, the groundwork.

VOL II – A solution for PI, just because Engineers CAN. More questions about the Universe.

VOL III – ENOUGH questions. Tying all things together and why. A Unified version of the Universe.

VOL IV – Completing the Unified Theory in more detail, especially Gravity. Last Book, UNLESS....

VOL V – Questions from readers. A Clarification.

Dedication Page

After Unifying the Universe there are many people that I need to thank, because let's face it, I never could have done this alone. I'm not that smart!

TO,

RJ, my research assistant, my sanity check, my Son.

Danielle, Thanks for all the Stars. My Star!

Amandá, Thanks for the Brain Trust, a place to think and work, and the courage to move away from home.

Rachael, Thanks for the 5K's! Or a 100 other places we've done.

Jake, it all starts and ends with you. You are my Universe. Someday you may even stop looking at me funny!

Zoe and Millie, My Brain Trust. Your wonder of what makes things move (treats) in the Universe is Amazing.

And to all the readers who asked questions and made me dig a little deeper.

COPYRIGHT PAGE

TABLE OF CONTENTS

INTRODUCTION

This book was not supposed to be. I had my Views on the Universe and it took Four Books to give my take. But then something happened that I never anticipated People would ask questions about my views. And since I included my Email on the Copyright page, it was sort of inevitable. And if some people had questions, maybe a lot of people were wondering. Was I clear enough in my Chapters? Then it became apparent that if I was going to answer questions for people writing in, I should just publish the questions and answers. This will not be full blown explanations of the Subjects for the most part, but will be an attempt to answer the question and be clear.

And to be clear, I am an Engineer not a Physicist or Mathematician. I try not to use Math to Prove or explain my ideas, just plain words. And as for Physics, well I don't want to say it is a mess, but there are so many "Theories" floating around and they often conflict with each other. There must be 50 Big Bang Theories alone. How can they all be true? Aren't Theories supposed to be a proven Hypothesis? So, for the most part, I just give my Engineering view. And if it agrees with Physics great, but if it doesn't, I at least try to provide any facts. Facts seem to get lost in the Universe. And by that I mean, to absolutely prove something, you list all the Physical characteristics then design tests to verify each and every characteristic, keeping in mind that there may be other ways that can give the same readings; The tests must be very specific. Then you claim Victory and you have the Truth. The subject that started me writing was the Big Bang. One version hypothesized that a big explosion should produce Radiation. When the Background Radiation was discovered, some claimed Victory and reacted as though this proved the entire Theory. Not even close! There is nothing in the Universe that is even remotely like the Big Bang Theories. If we are going to speculate, at least it should be similar to existing processes in the Universe. There is no compression anywhere that could compress the Universe down to the size of a marble. The point is; so much speculation, so little facts or even circumstantial evidence.

And as an Engineer commenting on the Universe, I have been accused of being straight out of the 1800's, where Reality is only things that can be sensed. Or in other words, the hear-see-touch Method. But I don't mind, since there are about six million Engineers working today (Bureau of Labor Statistics) vs Physicists(less than 50,000) And amazingly, more Engineers have flown in Space than Physicists by a wide margin! Why wouldn't we comment on the Universe. I think those 1800's values have served me well: Physical Proof, Physical Proof, and Physical Proof. Physicists work very hard, but. In my estimation, tend not to look at the big picture. Engineers don't look at events and say "Why is this happening?". No, we are likely to say "HOW is this happening." We look at the mechanics of the motion: how is it happening and where have I seen this before? Looking for commonality, not uniqueness. When it comes to motion, Engineers have a distinct advantage. And I will give you an easy example: When it was discovered that Galaxies were moving away faster than Light, Physicists defended the nothing can go faster than the Speed of Light Barrier by claiming that the Galaxies weren't traveling faster than Light, the Universe was Expanding. I laughed! Let's look at this from an Engineering View. If you sit on a Horse and it walks 10 MPH, how fast are you moving? 10 MPH! And if you board a passenger jet that flies 500 MPH across the Country, how fast are you traveling? 500MPH! You people are smart! Now, if I am I sitting in a Universe that is expanding Faster than the Speed of Light, how fast am I going? On the back of a walking horse or the back of an expanding Universe, speed is speed. Speed doesn't ask the propulsion system. This is so simple to an Engineer, but Physicists don't get it, because they have been taught that Nothing can go faster than the Speed of Light. An untested Einstein Theory, by the way. If it doesn't make sense to an Engineer, I write about it! And by that, I mean that I look at the Mechanics. Sure, an Electromagnetic Wave, that is what Light is, may not be able to go Faster than Light, BUT a Mass Object, like a rocket ship or Galaxy, might be able to. Different propulsion systems for Light vs Mass Objects, so likely different Speed Limits for both. BUT, BUT EINSTEIN SAID? Well Einstein may have known Light, but Engineers build Spaceships. When we run out of Ideas and can't go Faster than Light, we will let you know!

Here are some Questions and my answers to the Universe. This has been fun, but I think I will go back to making Cars.

WHAT THE. . . . .

I love the Universe and I love Physics :), ok maybe not so much, but every time that I think that I am done with this series of books, somebody asks; well what about this or what about that and how does it fit into the Universe. I'm pretty sure that it is God testing me to see if I have learned anything at all about his Universe. And here is the difficult part; Physicists are unable to Unify the Universe and what that means is that they are able to write about individual topics and never show how it affects anything else. When have you seen an article on how Sub-atomic Particles affect Black Holes or about how Black Holes affect the formation of Stars, etc. There are NONE! Why? Because each Subject is its own little world, there is no link to the rest of the Universe and there is no requirement to try. I, on the other hand, do have a Unified Theory that shows common links in the Universe. I am sorry that when I start talking about BlackHoles that it may lead to writing about the Big Bang Theories. But this is the way it should be; every subject in the Universe is related to the whole. Or as God thought, "I can make this so complicated that not even Einstein can figure it out or I can make it so simple an Engineer can do it." Ok maybe that wasn't an exact quote, but you get the point. So, Ok, a few more questions and I quit!

NUMBER ONE QUESTION

Since you build the universe from just one Particle, an Electron, does that mean there is only 1 Force?

This is very perceptive and also very True. God would be proud. Oh, I bet you want to know what it is, right. Okay, the answer is pretty obvious. It is Heat. I have stated before that I thought Thermal Dynamics should be a Fundamental Force, but is really just Heat. Why? Okay you know the drill; Mechanics!. Physicists know that as you near absolute Zero, the movement of the Electron become less and less. What they don't seem to realize is that there is and end point. A point at which all movement stops. Now Physicists like to give some mumbo-jumbo Quantum crap, but the fact is there is a point where the Electron will stop or freeze. That is simple logic. Just because we can't achieve Absolute Zero with our tiny brains, that does not mean it is not True. How come Einstein did thought experiments and was praised by the Physics community, but this simple thought is held as not true. Whatever, it is. And, all movement, all Forces cease to exist. No Strong Force, Weak Force, Gravity, Electromagnetic Force, Nothing! No momentum, No centrifugal Force, Nothing! But Remember, THIS IS JUST A THOUGHT EXPERIMENT! SO, HA,HA, JUST KIDDING. Energy can never be created or destroyed, Only transferred. There must be something there to transfer the last bit if energy to. But even then the two things can only reach equilibrium, so the last bit of heat gets shared. OK, if Heat is the one true Force in the Universe, where do the others come from? Well, just like everything else in the Universe, the other Forces are built from the ground up. What does that mean? As an Electron heats up, it begins to spin. All of a sudden or slowly we have a charge. Enough charge, we have Gravity. Enough Gravity can cause some Electrons to reverse spin. Positive and Negative Electrons can build Protons and Neutrons. And on and on until we have all Forces and all Matter. We have a Universe. All we need is a little Heat. And just to be clear, this doesn't mean that the other Forces are not Forces, but it all starts with heat, The real fundamental Force. Now not to confuse you, but I have already said that Heat(Energy) can not be created or destroyed. But it can be diluted, which has the effect of reducing it. In the broad sense, Entropy. So when you see a claim of reduction of heat to below Background Radiation levels, 2.7 K, that is very hard to believe. Notice I did not say impossible, I just said hard to believe. Yes, there could be a localized cold spot artificially created but, remember, that to offset a temp of 2.7 K, there must be an area less than 2.7 K. And if the claim is a Temp. of a Billionth of a Degree K, the localized area must be less than a Billionth to cancel out the ambient Temp. Not impossible, but very unlikely. Along with that, there are also the things that I have mentioned earlier. Namely the collapse of the Atom. The electrons simply would not have enough energy(Weak Force) to continue circling the Nucleus and charge energy, + and -, would take over and the Electrons would bond with the Proton(known as a Strong Force) Yes, in my Unified Model, that is exactly how a Neutron bonds to a Proton as well, only the charges are slightly less. Interestingly, that electrical bond between a Neutron(-.5), an Electron(-1) and a Proton(+1.5) could last all the way to Absolute Zero, when the heat finally runs out or absolute Entropy. So the Atom requires Absolute Zero to drift apart, but the collapse of the Atom should occur well before that, at say a Billionth of a degree K. Again, to achieve Absolute Zero, the area around it would have to be 0 K. as well, so that the heat from the Atom can transfer away.

So, to recap, Heat is the true Fundamental Force, without it there are no other Forces. After that. Electricity(Electro), Gravity, Magnetism. That's it.

THE MECHANICS OF LIGHT

In the previous Books, I tried to give you the concepts of how things moved in the Universe. It occurred to me that most times I was speaking in Engineering terms, but most readers were probably not Engineers. I am making an effort to explain a little clearer in plain English what the mechanics are or How the Heck do Things Move and React; Cause and Effect. And since Electromagnetic Waves do most of the movement, it is a good place to start. If I had my way, I'd start with James Clerk Maxwell, for two reasons. 1. Did he think there was an Electron in Electromagnetic waves. 2. Did he realize that his use of calculations to predict would dominate the field of Physics

Now, it all starts with the Electron. The Electron is the energy carrying particle in the Universe. In fact, it is the only particle in the Universe!, but a different chapter Now before all the Physicists start complaining that this is wrong; here is what it can do. Being the smallest of the Fundamental Particles; an Electron, a Proton and a Neutron. I can use Electrons to build a Proton and Build a Neutron. And once that is done, I then can use those three particle to build Atoms, Molecules, Elements, Planets, Suns and the Universe! But what about the Fundamental Forces or all Forces in the Universe? Well, guess what, an Electron can do all that too. Everyone already knows that Electrons carry a Negative charge, but what is not known is that Electrons normally spin in one direction to give it a

Negative charge, but it Stars under tremendous heat and pressure, some electron can actually spin in the opposite direction to produce Positive Electrons. In the cooling process, the Positive and Negative Electrons can combine to form Protons and Neutrons. Pretty simple. But what about the Forces? Again, there is agreement that mass objects carry Gravity. No doubt Electrons have Mass, so Check that box. Electrons can be present in Electromagnetic Waves, in a minute I will show the mechanics, So Check that box. Electrons make up Protons and Neutrons. So Strong Force, Check that box. Electrons are in an atom circling the Nucleus, So check to the Weak Force! See how easy this can be. But back to the Mechanics of Light. So electrons are the carrier of energy in the Universe, of course, since they are the only particle in the Universe. Alright back on Mechanics. An electron with no Energy just sits there like a bump on a log. Much like a sports ball, same thing, put energy in it by kicking, or throwing, or hitting and the energy is transferred to the ball and it can move. An Electron is no different; when you put energy into it, it begins to move, up and down, or vibrate. I have another Propulsion Theory that Heat is the only Propulsion Force in the Universe, but back to Electrons. Most Physicists believe that the horizontal movement of this vibration (a wave) figures into the frequency. Nothing could be further from the Truth. An Electron just sitting in one spot, vibrating up and down so many times per second, would have the same frequency as a particle moving horizontally at the speed of Light, but only going up and down at the same rate as the one sitting there. So what does this all mean. Light is a Frequency that our eyes perceives as Light. But it would not matter to our eyes if it were travelling 1 mph or 100 mph or 1 million mph. It would all be Light to us.

Think of a Ball bouncing at a frequency. The Ball continues bouncing but is propelled horizontally at the Speed of Light. To us, the ball would appear to be moving as a wave, but the ball is really just bouncing up and down, while it is propelled horizontally. Duality of Light? I call this the "Can't Walk and Chew Gum Theory!" Now I am going to conveniently avoid the discussion of Photons, because Photons don't exist, only

 Electrons. In fact, Photons are one of the main reasons that NO PHYSICIST has been able to Unify the Universe. I, an Engineer, can and guess what, no stinking Photons. My Universe is elegant, but simple. I truly believe that the Photon was invented, that's right

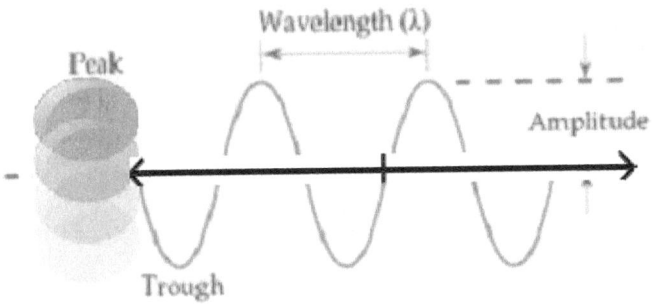

INVENTED, because Einstein Theorized that no mass object could go the Speed of Light, because its mass would increase.

So the Electron had to be replaced with a massless Particle. Right, and Pigs can fly!

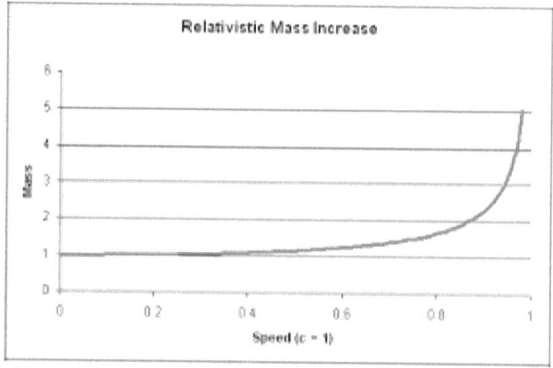

Oh, Oh, I can hear the question coming, Well if you are so smart, Mr. Engineer, why don't we see Light traveling at 10 or 100 or 1 million mph?? Why is it only traveling at 186,000 mps? Excellent question. This is a separate Chapter again, because Physicists really have a problem with this obvious Concept. But since I am not a Physicist, I will give you the Truth. Think about what Electro means in Electromagnetic Waves. Good guess, it means something to do with Electricity. Now how does electricity travel from point A to point B. Right again! It needs some kind of conductor, like copper wires, But under certain conditions several things can conduct Electricity. So, on Earth, the speed on the conducting medium, like copper wire, has been measured. And lo and behold it was less than the speed of Light. But my textbook says all electromagnetic waves travel at the Speed of Light. So it was tried again in the air, the water, and glass. And again the conducting medium changed the speed! And they were all different!. So, an Engineer would logically say, there might be something in Space that limits the Speed of Light. That is what the hard evidence is suggesting. So I know this was the long way around to answer the question, Why does Light travel at 186,000 mps in a vacuum? Physicists have been fighting this for a hundred years, the concept a few thousand. There is an Ether, even Einstein said Gravity can distort Space, therefore there must be something in Space to distort. And at the risk of being shunned by the Physicists, it is called an Ether. Want some more proof? Next Chapter.

THE ETHER

No one really asks me about The Ether, they just ask why I believe in one. And I usually say because all signs point to it. And that leads to a long discussion and more questions. So, I have been alluding to the Ether in every Book, but maybe I haven't made it very clear or explained it well enough. My bad, and I will try to say it better. We will start at the beginning; if you go outside and yell, somebody two blocks away may hear you yelling. That is because the Air or atmosphere is made of atoms that can carry your voice vibrations. And to prove that, if you were in a vacuum or the vacuum of Space, no one would hear you, because there are no atoms to carry your voice vibrations. And on and on about how dense things can carry sound much better, because there are more atoms. The fact is that your voice travels in a wave, it's called a sound wave. Now we come to electromagnetic waves. This works just like your voice, except your voice is mechanically generated and electromagnetic waves are electrically generated. But the same rules apply. There must be something to carry the wave and it must be a conductor. Just as your voice would not carry in a vacuum, Electromagnetic waves would not carry in something that is not a conductor. Electricity is carried very well by copper wire, but hardly at all by a rubber or plastic cover. No we come to Radio and TV. They discovered very early that radio, then TV waves carried pretty far in our Atmosphere. It was apparent that our Atmosphere was a conductor. Finally, we go to Space and low and behold, Space is conducting Radio signals. But here is where things went horribly wrong. Some people including Physicists forgot about the conductor part. They forgot about sound waves, Lightening, and Radio and TV signals. They thought that Electromagnetic waves could travel thru Space, all by themselves or somehow by their own propulsion. I have said in every Book that every Electromagnetic wave needs a conductor and the best conductor would be Electrons, but Atoms could do that, too. That means All Space would have some kind of organized matrix of particles and it would be called the Ether. Now I told you that all signs point to an Ether, and here are the Signs:

1. Both Einstein's Gravity and Warp Drives work by distorting Space. What do they distort if there is nothing in Space?

2. The Speed of Light is fixed at 186,000 mps. Why? Maybe the Ether limits the Speed. Why does the atmosphere, glass, and water further limit the speed.

3. Space has been measured to have a Temperature of 2.7 K. If there is nothing in it, why not 0 K. Maybe an Ether has heat, or at the very least conducts heat, as an electromagnetic wave.

4. There is missing Mass and Energy in the Universe, Dark Matter/Dark Energy. An Ether could make this up.

5. The CMB, Cosmic Background Radiation could actually be the Ether or part of it.

6. Both Light and Gravity can travel across the Universe, both are electromagnetic waves. Maybe they are carried by the Ether.

7. Astrologists have long thought that the early Universe was filled with evenly distributed Hydrogen and Helium Atoms for Approximately the first million years or so. Some would condense into Stars, planets and Galaxies. If only some, where did the rest go? It would have continued to expand and form the Ether and provide minimum heat it deep Space.

Seven significant signs of an Ether. This should be a No Brainer. But sometimes Einstein was wrong and didn't think it thru. He thought a Cosmological Constant was his biggest mistake, but in reality it was not supporting the Ether, but then that would have negated SpaceTime. But then he had to give Space the qualities of the Ether; made of Mass particles that can be deformed by Gravity. Coincidence, I think not. Especially since Einstein was aware of the Michelson-Morley experiment. And since this supposedly proved the Ether did not exist, this opened the door for SpaceTime. Einstein conveniently stayed away from defining what Space is made of. Was this an oversite or a plan to cover the fact that he was just combining the Ether with Time? The amazing part is that no one ever called him on it. I am sorry for writing about the Michelson-Morley experiment, but this is what proved there was no ether. It was thought for a long time that the Ether was a matrix in Space. Well Michelson-Morley thought that if they measured light moving with the Earth through the matrix, it would move slightly slower than Light moving perpendicular to the movement of the Earth. They devised a way to measure both ways. Needless to say, The measurements were the same and the Ether concept was throw out by most. This was completely WRONG and here is why. You never assume physical properties for something you are testing. You devise a test to prove those properties, then build on that. They could have been right, but they equally could have been wrong. We just don't know. So the experiment was wrong for two reasons:

1. There was no proof the Ether could slow down or speed up Light. True the Earth would be moving thru the Ether, but we know that speed is not additive or subtractive to Light. Light merely moves at a constant Speed, which is exactly what Michelson-Morley measured, no difference in Speed.

2. There was no proof that the Ether even effected the Speed of Light. Michelson and Morley just ASSUMED it did. And when Light failed this Assumption, Michelson and Morley did not question the validity of their Assumption, even though the Assumption was just made up or not based on some Proof.

So, the Ether was thrown out with no Physical proof. But yet Einstein claimed that Gravity can distort Space. Really? What

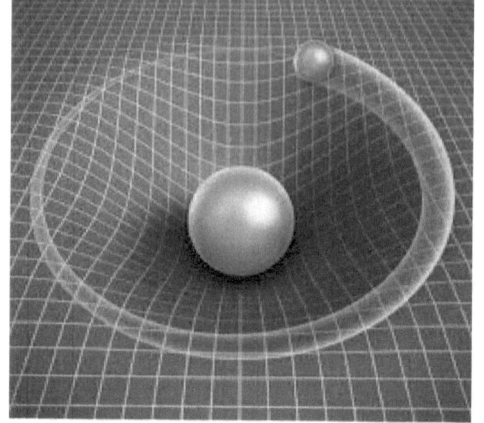

could there be in Space that can be distorted?? And the true irony is that SpaceTime is pictured as a matrix in Space: What is

this fabric of Space that is pictured? Yes, Michelson-Morley, Light can go equally fast in any direction, because the Ether is really like a Matrix not a river. You're assumptions were wrong, therefore your experiment was INVALID. And even this picture is wrong, as Space is 3D not 2D. Space should look like this:

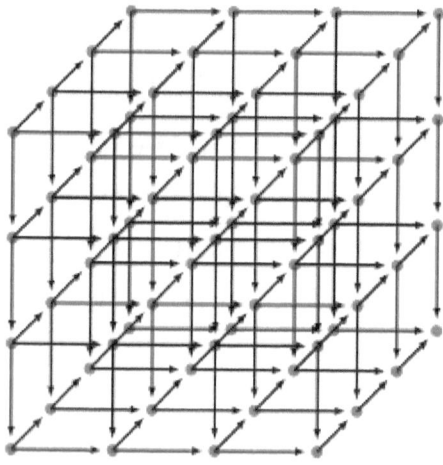

Enough proof? I believe so, because a 3D Ether can be distorted! And one day the Scientific Community will come to the realization that only Electromagnetic waves are limited by this Matrix to the Speed of Light, Mass objects just push right thru it; the way Planets, Stars and Galaxies do today. Galaxies moving away faster than the Speed of Light? No surprise to me. Cracks in Physics in the late 1800's? Here is one today; So much time spent on how electromagnetic waves move, but very little on how Mass objects move. They are different animals you know. And both Theories are wrong. Fix it!

THE BIG BANG MECHANICS

This is not intended to be a rehash of the Big Bang Theories because they just hurt my head. This Chapter is more about why these Theories can't exist, even within normal Laws of Physics. To explain, matter has physical properties: In Engineering terms; length, height, width, size, mass, energy levels, etc. For most Big Bang Theories it is predicted that these properties will be changed in some way. Most times simple compression would seem to be a solution, but again,. most Theories just can't let it be. Even speculating compression ratios well beyond Black Holes or anything seem in Nature. In fact, there is NO physical evidence that a Big Bang of this kind could ever happen! Why? Because we know that even Black Holes radiate Energy.

Imagine taking a Black Hole and compressing it down to the size of a Pea. Energy would just be pouring from this compression. By the time you got to a Pea size, most of the energy would be long gone. BUT, BUT, What about an Explosion? Yes, that is highly likely. As matter is compressed it will start heating up. And since the Universe is mostly Hydrogen, at some point well before a Pea size the Universe will explode. No need for a break down in forces, because as soon as Hydrogen gets hot enough, the Universe will explode. A compression ratio down to the size of a Galaxy would be enough to trigger a massive explosion. BUT, BUT, what about those little particles that instantly appear and disappear? Couldn't enough of those particles appear all at

once and cause a Big Bang? Well in the minds of some
Physicists it could, but the mechanics just doesn't support that.
ALL particles would have to appear in a relatively micro short
period. But what Force would cause the compression of trillions
of particles all at once at the same time? So the conclusion
based on the mechanics is NO, these Big Bangs have NO
chance. Especially when they are compared and contrasted to a
Supernova style explosion. Make these Big Bang people go
away. Science Fiction is nice and interesting, but if there is no
Physical evidence that can support a Theory, then it's most likely
wrong. And if there is nothing in Nature(The Universe) that is
similar, then it is most likely wrong. Black Holes are assumed to
have infinite Density, yet they don't explode. You can't get
more dense than infinite, so how is a Big Bang going to explode?
Or even exist? The logic must be consistent. And speaking of
consistent, what is the CMB really? If there was an explosion
shouldn't there be a spherical CMB? Basically there is! But the
the Universe is expanding in one direction? Then shouldn't the
CMB distribution reflect that? So many conflicting questions
and almost zero evidence. It would make more sense to
speculate that a Giant Black Hole exploded. Or Not! And I say
Not because because as with all Big Bang Theories and Black
Hole Theories, they start with a Sphere, but not just a sphere, a
compressed Sphere! That is significant in that the higher the
compression, the more likely the contents of the Sphere will

become homogenous. And what this really means is that from a mechanics (an Engineering) standpoint, you can't get from Here:

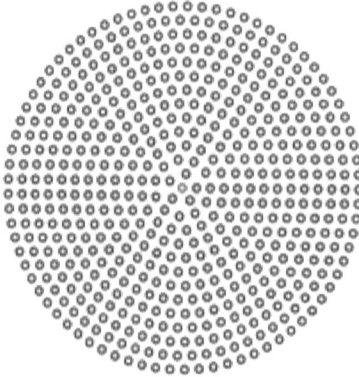

To Here:

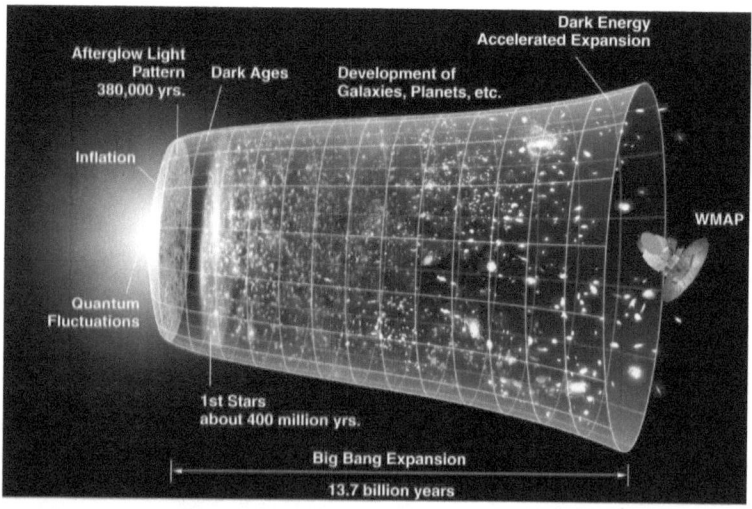

This indicates the SPHERE was Not highly compressed and homogenious, but more loosely packed and allowing for a directional explosion. Case Closed!

MATHEMATICS

It seems as though the only useful place for Math is in Physics. We know this is not correct, but Math takes on a special place in Physics. In the general sense, Math is just shorthand for working with numbers. Although it can be done, high level languages are not very well equipped to handle very large or very small numbers. Add to that complex formulas and equations and it's easy to see that Math is its own language. But what gets lost is that just like any other Language: Garbage in Garbage out. Too many people believe that Math is Proof. So as an example, 1 + 1=2, but 1 apple + 1 orange = 2 objects and 1 apple + 1 nuclear bomb = 2 objects. So what did we actually Prove? Nothing! What really does the proving is the assumptions. I have all apples or I have objects. Math can only do what it is told. It can't think, it can't predict, it can't anticipate. Remember the old saying, "Numbers don't Lie, but you can Lie with numbers". What does all this mean? Math results or the calculations can be biased by the inputs(assumptions). Yes the Math may be flawless, but the assumptions could be a mess. So be very skeptical of results that are only supported by Math. Sure, Space is very large and Electrons are very small, but look how the perception of the Universe changed radically when Telescopes were invented. The Gold Standard always has been and always will be, physical proof or testing. Calculations are interesting and seem so accurate. But Math is just a language similar to

English, French, German, Chinese, etc. Great stories can be told, so real. But always know the real Truth or Proof is Physical. I was going to stop there but I thought of a few more examples. Ever heard of the Planck Length? This is thought to be the smallest length anything can be. But since numbers go on to infinity, why have a limit? Just use a number. This is odd considering on the plus side, infinity is often used. And on the plus side we can only see about 14 billion Lightyears. So there you have it, a Planck Length and a Planck Distance. Or as I, an Engineer, like to say, who cares! And the final example of great Math would be the Geocentric Model that shows the Earth at the center of our Solar System. This was an accepted view for over a thousand years. The movements of the Planets and Sun around the Earth was eventually calculated fairly accurately and the positions could be predicted. The Math was good, but the assumption was wrong. The Planets and Sun did not Orbit the Earth, even though the Math "Proved" it. Be very careful with Math.

And since we are talking about math, let's use PI as an example . . . . .

PI

This example shows how the math can be correct, but the logic is wrong. Let's pick on Pythagoras and his Theorem. His Theorem is a thing of beauty, but it is wrong. For a right triangle, $A^2+B^2=C^2$. Here is a pic,

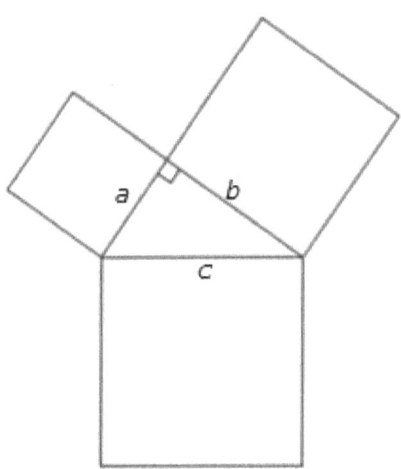

Now, what is wrong? It's ok, nobody has got it yet, I will tell you. **A** is a definite length, **B** is a definite length and **C** is a definite length, So, in the formula $A^2+B^2=C^2$, C^2 can never be an irrational number because we know that C is an exact dimension. BUT that is exactly what C can be per this formula. As an example, if A=1 and B=1 then C= the square root of 2. But that is an irrational number, therefore the answer is wrong, because A, B, C are cubes with exact dimensions. Now you are probably thinking, what the heck does this have to do with PI? Sorry to take the long way around but the same flaw is in the formula for PI. And what does that mean? Again, multiplying the diameter by PI should never produce an irrational number. That needs to be corrected along with the Pythagoras formula and every other Greek formula which includes Archimedes and Euclid. See the problem came about because Archimedes tried to use multisided polygons to calculate PI, but that required the use of triangle calculations, which is where the flaw is. So, it really wasn't Archimedes fault, it was some other Greek who developed the formulas for triangles, possibly Euclid. Now I can tell you what is wrong, but as an Engineer, my Union contract doesn't permit it. But I will tell you that PI is **3.15**, take it to the bank!

So, what was this for? This was just to show that math can show whatever story you want to tell, if you are clever enough or not precise enough. Be very careful with math.

CMB

You know, I don't even know what this stands for, had to look it up. But I do know that it may be the missing Link. So, again, let's look at the Mechanics. First, I think Background Radiation is a misnomer. If the Universe were a giant Sphere of explosive particles and gases, it would seem that this wall would be in the forefront of the explosion, not the background. Now assuming that the blast was of the Nuclear variety, which, surprise surprise, creates radiation, then one could assume that everything in the Universe was vaporized. We would have one big cloud of tiny particles. And what are tiny particles called? That's right, Electrons. And what is Beta Radiation made of? Right again! Electrons! So, apparently, this cloud of Super heated Gas begins expanding very rapidly, But as it expands it begins to cool. When sufficiently cool, the particles begin to interact and bond. And since this chapter is as Science Fictiony as I get, some particles spin one direction and some the other, thereby creating opposite charges. This leads to the formation of all kinds of Mass objects. And the slower and cooler the Universe gets, the bigger the Mass objects get. This eventually leads to Stars and Planets and Galaxies. Well that is the short version of the Universe, but what is forgotten in all this commotion is the rest of the particles that continue expanding. Now, since everything was expanding from the explosion center point, the Cloud of Particles and the newly formed Mass Objects are riding the expansion together. Although millions to billions of Mass Objects form, it is only a small percentage of the Total, let's say abought 20%. Hey, wait, isn't that just like observable Mass Objects VS Dark Energy/Dark Matter?? Calm down, it gets better! For Thousands of years People have believed there was something magical in Space. Something that could conduct Electromagnetic waves and they named it The Eather(Ether). Now could it be just possible that the CMB has proved the existence of the Eather? Let's look at the facts:

1. The Ether was thought to be everywhere in Space – The CMB shows that radiation is fairly evenly spread across the Universe.

2. It is thought that (by me) that Electromagnetic Waves need a conductor to travel along – CMB radiation can be Electrons, which make a great conductor.

3. Einstein said Gravity can distort SpaceTime – Electrons are Mass Objects and therefore, if SpaceTime is made of Electrons it can be attracted(distorted) by Gravity

I'm glad that this is my Science Fiction Chapter, because in real life I would be inclined to say that the CMB helped to prove the Ether.

EINSTEIN'S GRAVITY

Gosh I hate this Subject. I can't believe anyone actually thinks this is correct. I laughed when I first read it, but then I realized this was being taught in our Universities. Ok, I won't bore you with my opinion, as an Engineer I will give you facts. This is a common picture that illustrates Einstein Gravity.

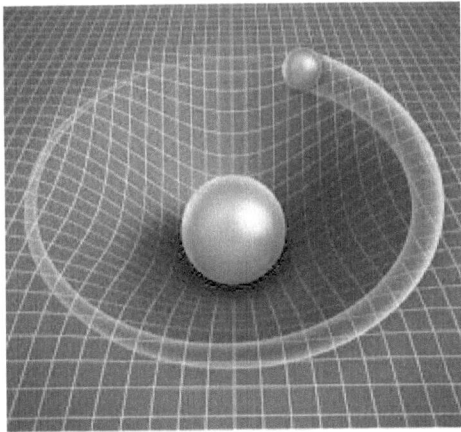

Let's call this the Sun and a Planet. Now the Theory is; The Sun distorts SpaceTime and the Planet merely follows the distortion around the Sun. And people agree. Well for starters. I call this the Bowling Ball on a Trampoline picture. I say this is laughable and I will list the number of things wrong.

1. I was taught that Gravity is an attractive Force. This has been proven a million times over. Now look at this picture. It appears as though the Sun is actually repelling SpaceTime! If the Sun actually attracted SpaceTime, it would be wrapped around the Sun, being pulled right to the surface just like Gravity on Earth.

And to support that, objects can be 10 miles above the Earth and still can be pulled to Earth. It works spherically everywhere around the Earth. Now squint real hard and try to picture this picture spherically, What a mess! Are you laughing yet? Well it gets funnier.

2. The distortion zone is circular or spherically around the Sun. That means Planets should only have circular orbits. Well guess what? Many Planets and other Mass Objects do not have circular orbits. What about Comets and Asteroids that loop around the Sun and head back into Space? How do they get in and how do they get out when Planets can't escape? Einstein's Gravity only allows circular. Oh no, what are we going to do! Well, before we decide, let's look at the more hilarious things about this Theory.

3. Take a look at the Planet, the only thing guiding it thru Space is the distortion of SpaceTime. Now picture the Earth, 5,973,600,000,000,000,000,000,000 Kg traveling at 67,000 mph. Imagine the incredible centrifugal Force being generated. So here comes the Earth like a Mack Truck headed right at that puny Spacetime wall. Is it going to stop the Earth from smashing right thru it and fly off into Space? What is it made of Titanium!

4. Now look again at the picture, Maybe you could visualize this working with one Planet, but Oh oh, the Sun has 9 Planets (I just put Pluto back for having to explain this) According to

Einstein's distortion, every Planet would have to follow one orbit. But we know every Planet has a different distance from the Sun. The only way this is possible per Einstein's distortion would be for faster Planets to be near the top and slower near the bottom all based on centrifugal force. So Mercury should be the furthest Planet from the Sun! But thankfully this is not that way. In fact, we have basically have a flat Solar System, a fact! The only way Einstein's Gravity allows for that is if all Planets follow roughly the same orbit or Gravity distortion.

5. Newton, remember him?, postulated 3 Laws of motion. The first being that mass attracts mass. That attraction being call Gravity. The next 2 Laws talk about how Size and distance can effect this attraction and by how much. Actually some pretty detailed calculations considering this was the 1700's. Einstein pretty much disputed these Laws in part, and let's be fair, he was explaining the bending of Light. The problem is that Einstein's Gravity attracts mass one way and bends Light another! And this is where it gets a little confusing, because sometimes Einstein's Gravity is thought of the grand Theory with Newton's Gravity being a subset. So for mass objects, Newton's Gravity is fine, but for electromagnetic waves Einstein is right. Well, you could believe that, but Einstein provides no details of how exactly that works. Newton provide an explanation of exactly how his Gravity works; Mass attracts mass, and based on the size and distance, how much Gravity there is. Einstein, on the other hand, merely states that Gravity distorts SpaceTime. No

explanation of how or how much this works. In fact, if you look at pics illustrating Einstein's Gravity, it looks like Gravity is repelling SpaceTime. And, of course, the ultimate question; WHAT THE HELL IS IN SPACE THAT CAN BE DISTORTED!?

Right after you quit laughing, think about how not one Physicist has ever disagreed with this laughable concept. I am so glad that in my Universe, Gravity follows Newton's Law and Light has a mass particle in it and we don't need Einstein's distortion to bend Light. Or if you prefer, Space has Mass in it (Ether)

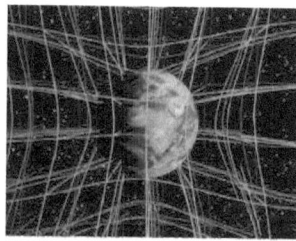

Gravity could distort Space if it had mass in it and therefore bend Light.

Newton's Gravity works just fine thank you. And before we end the Einstein Fun and Games, Imagine the Mack Truck Earth Orbiting around the Sun and the in-creditable amount of Gravity it must take to hold it in orbit. So cut the crap about Gravity being a weak Force. Let's make the Nucleus of an atom as big as the Sun and we will see how strong the Strong Force really is.

As unbelievable as this Chapter is, it is only one of the Big Three, this one plus the next two, as to why the Universe can not be Unified by Physicists. Luckily, I am not a Physicist so don't worry, I am used to making things work correctly. Can't wait to get to Warp Drives.

PARTICLE PHYSICS

This is a hard subject for me. I appreciate how hard they work, but I have to tell them that they got it wrong. In life, in Engineering, in almost everything, you start at the bottom and build up. So for Particle Physics what does that mean? Well, It seems as though they know what the next level up is supposed to be, so they tailor their discoveries to match that. In other words; I have a big particle, now what pieces do I need that can make that Big particle, Now I test and what do you know, I found or discovered the exact pieces that I said I needed. A Miracle. That is almost opposite of Science where we discover pieces first and then find out how they go together and why. Science doesn't normally work where you know what you need. That is not the Universe by discover, that is a pre-planned Universe. Why not just go to the store and buy one! So, almost from the start they knew that Electrons were one of the smallest, if not the smallest particle. You now have a small particle, but can it do everything that needs to be done. Let's say I need an Up Quark, can an Electron do that job? Well no, an Up Quark needs to be 3 times bigger than an Electron and have a Positive charge. Well what if Positive Electrons Existed? And what if we took 2 positive electrons, added 1 negative Electron. We would now have a particle that could do the job of an Up Quark, but was made of Electrons. You continue that process until we absolutely HAVE to introduce a new Particle. Things like a

Higgs Boson may not be needed because an Electron has Gravity(Mass). Keep it Simple. It is not that the current particle model can't work, it's the fact that we don't need it! One particle, an Electron, can do everything that 30 particles can do and more. The only thing needed is the Electron spinning one direction to have a Negative charge (normal state) and sometimes spinning the opposite way to form a Positive charge. That's It! With these two you can build Protons, Neutrons, Atoms, Elements, Molecules, Compounds, THE UNIVERSE! Sorry Particle Physicists, NO Higgs Boson. And yes, I go into more detail in previous Books, but the concept is always the same, Occam's Razor.

QUANTUM PHYSICS

Where to start? It's easy to criticize and say things like do we live in a Matrix and are we just Simulations? Or are we analog with a degree of randomness? I have deliberately stayed away from this discussion, because it may be just a Theoretical discussion. What do I mean? If you state I went to the Moon, the important concept is I was here on Earth and I traveled and ended up at the Moon. Of course I added that I started on Earth and I traveled. So here is where Quantum anything kicks in. The Quantum Theory leads to the inevitable conclusion that everything can be Quantified. The problem with this is right in front of everyone's face, This is the Old Half of a Half of a Half Theory. Even if you break things down into smaller and smaller pieces, Something can always be divided in two! For example. Take PI. Two billion digits and no end in sight. So Quantum Mechanics/Physics gives the illusion of more precision, but it is a fools game you can't win. Sometimes it is just good enough to say I went to the Moon. It might be interesting to know how long it took, how long I was gone, etc., but bear in mind that there is a point that infinite answers have no bearing or properly, infinitesimal bearing. So what is my opinion of Quantum Anything? Well just like this Chapter, a big waste of time!. If I am going to the Moon, I am more concerned about Safety, Redundancy, Eating, Breathing, Communicating, etc. than I am about a quanta of energy in Space. To me, Quantum Physics is just one big Thought Experiment. I saw a video the other day stating that the Universe is built on Quantum fields. No, because Quanta is just a made up word, as is Quantum Physics. And now you have just made up a Quanta Universe. I prefer the real World. I prefer the Randomness. But so far this Answer to a question has just been one big Rank. Here is the question asked; Is the Double Slit experiment real? First, I hate commenting on things that don't advance Technology or Humanity. Second, I would rather talk about how Pigs can fly. But since you asked, the answer is NO. God, I wish I could stop there, but I will give you the mechanics of the experiment. An Engineer would run this a little(a lot) differently. First, if we have an Electron Gun capable of firing ONE Electron at a time, we can test the repeatability of the Gun. How? We fire Electrons at the detector without the double slit screen. If the shot pattern is a very tight circle, we are ready to move on to the Experiment. But if the shot pattern is wildly inaccurate, we need to correct this in accuracy. This could mean anything from replacing the Gun to placing a tube over the Emitter barrel that extents all the way to the target, thereby not allowing any stray shots. I can say with 100% certainty that if you do this, all Electrons will go thru the slit the Gun is aimed at. But what if we aim at the center point between the Slits? Then, guess what? No Electrons will go thru either slit! To an Engineer, the Double Slit Experiment is a classic example of running a Test(Experiment) incorrectly. And Yes, I have seen times where they try to make the slits very accurate. But It's all about the accuracy of the Gun. I have yet to see this Test ran Precisely. And if you do take time to make the Gun accurate, then you can claim crazy things like Superposition, but if not, SHUT UP and watch my Pigs fly.

LESS THAN ABSOLUTE ZERO

I think the real story here is not what Absolute Zero is or even if we can achieve it. No! The real story here is how Space, empty isolated Space, can maintain a 2.7 Kelvin temperature. The first part of this Chapter I question if we are actually getting close to Absolute Zero, and then if that is true, then why is Space at 2.7 K. And if that is true, then WHY? And how this is just another case of Physicists working in their own little bubbles and no one is paying attention to the Whole or Big Picture. I am….So, I know very little about Absolute Zero, even less than I know about Physics Theories about the Universe. But this Subject seems to have more Science behind it. Lots of people have studied this Subject. As with everything, I study the Mechanics first to see how and why it moves. But I look at this and something just doesn't seem right. Let's see, where to start? Now the Universe or empty Space is thought to be at around 2.7 degrees Kelvin, But Scientists believe they have come to a billionth of a degree of Absolute Zero with a Helium Atom right here on warm Earth. That is very hard to believe, what could be colder than empty, isolated Space. And finally one of the techniques for cooling was Laser cooling and Magnetic control of the movement of the Atom. Both of which should add heat to the system. I'm not sure, but in the race to go colder it seems like someone is playing fast and loose. And more unbelievable was the fact that in one room it was almost Absolute Zero and in the next in was a comfy temp. for the Scientists! But lets go with that billionth. I think where I start doubting the process is in the Mechanics. Essentially, 2 Protons, 2 Neutrons, and 2 Electrons in the lowest energy state. The Mechanics are such that the Nucleus spins and this imparts centrifugal Force on the Electrons to keep them in orbit about the Nucleus, just like our Sun and Planets. Yes, pay attention Cloud of Electron people, a flat Solar System and a flat Atom, Really, this stuff is all related. Now there is a perfect balance between the Gravity of Nucleus and Electrons, plus the Positive charge of the Nucleus and the Negative Electron. Now because the Nucleus, (Proton and Neutron) are thousands of times larger than the Electron, the Electron must orbit the Nucleus very fast to have enough centrifugal Force to prevent being sucked in to the Nucleus. Now this where things can go wrong. The relationship between the spin of the Nucleus and the centrifugal Force or speed of the Electron cannot be simply Linear, because Gravity as well as charge that have to be overcome by the Electron. And to simplify; There is a minimum Escape Velocity speed to keep the Electron from being pulled into the Nucleus. And remember, the Electron is thousands of times smaller. And it so happens that this speed is well above a billionth of a degree mark of Absolute Zero. At this Temperature, the Nucleus would be rotating very slowly and the Electron would have almost no centrifugal Force or speed. No doubt the electron would be combined with the Nucleus and the charges would be cancelled out. Possible, but

this is not how they tell the story. And you would no longer
have a Helium Atom. You would only have six particles, two
Electrons, two Protons, and two Neutrons, barely held together
by Gravity. So either they didn't see what they thought they saw,
or Absolute Zero is less than they think it is. Can't be both. And
this brings up my favorite question; Why does there even have to
be an Absolute Zero? There is no way to measure it without
introducing heat, and without a way to measure it, all we are
going to hear are calculated wild claims. It serves no useful
purpose. Let's just throw it in the scrap heap with all that Planck
stuff and move on to something useful. And let's put an end to
this discussion with a few more facts; Space is empty, I mean
really, really empty. Now according to my calculations, which I
don't do, there is about 5 atoms per square mile in deep Space.
Even a million atoms per square mile would never be able to
keep Space at 2.7 Kelvin. Why, because the heat carrying
capability of Hydrogen, with a Proton and an Electron, is very
low. Even with Helium added to free Hydrogen in Space,
supporting 2.7 K would be impossible, as Helium is 10 times
less abundant than Hydrogen. So these numbers that people are
throwing around for Earth based record low temperatures are
absurd compared to empty Space. But there is a way to make the
numbers correlate. Now suppose Einstein's fabric of Space were
actually a 3D matrix of Hydrogen atoms, but more likely,
Electrons, evenly distributed throughout Space. This would help
to explain so much: with this huge quantity of particles in Space

2.7 K temperature could be maintained. As well as an

explanation for Dark Matter, Dark Energy, An Ether, and why

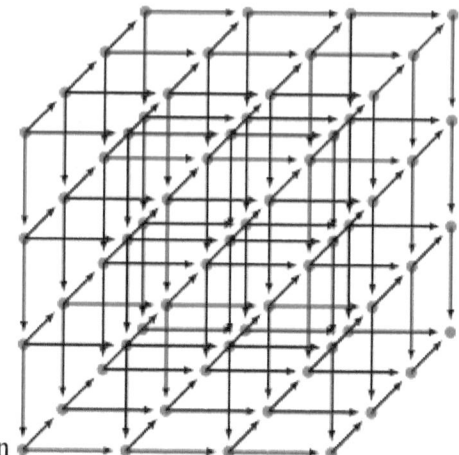

Einstein's Gravity can

distort Space. Of course this is

too much Big Picture for

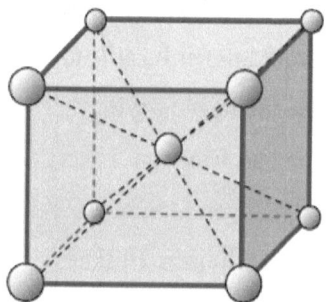

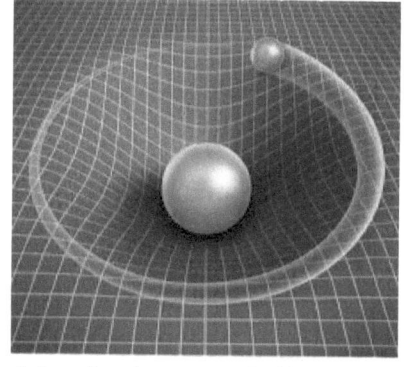

Physicists, but hey, we tried! .

SUPERMASSIVE BLACK HOLES

This title is a misnomer, as all Black Holes follow the same Laws, so this chapter is really about all Black Holes. I try to refrain from opinions, But Black Holes amuse me. It's not that I think that Black Holes can't exist, it is the definition that I disagree with. I always start out with the Mechanics and go from there. Everything in the Universe is basically the same process; Gravity pulls things together, even at small scales and at extremely large scales. This is nice, but when it comes to Black Holes and things are different. There are basically , 2 ways a Black Hole can form, Let's look at each:

1. A Star can burn most of the fuel from its core and become hollow enough to collapse inward and explode or go Super Nova. This explosion pressurizes the core and from a Black Hole, with the rest of the star being exploded away. And all that is left is the Black Hole. This is a great Theory, but the mechanics don't support it. If the Star collapses, fresh fuel will be pushed into the core and be just as likely to explode as anything else as it's at the core and under the most heat and pressure. At the least it would burn with the explosion, but since it is the hottest spot, it probably where the explosion occurs. No Black Hole.

2. Thru Gravity, a massive object begins forming at the center of a Galaxy. Again this is a nice story, but a couple of things

against it happening. Centrifugal Force generally would keep objects away from the center, so limited opportunities to form and get bigger, and the process of Gravity would be the same at the center of the Galaxy as in Space. If there was an object it could be a Star or more likely just a massive Planet. Possibly, if there were enough material available something could form, but here is the part that is not explained very well, why doesn't this happen everywhere? You could Theorize that Stars could happen and they do everywhere or Planets do exist and we see them everywhere, but Black Holes? Not so much. And the real thing that bothers me is the fact that Black Holes have massive Gravity; So why aren't all the Stars in the Galaxy in the Galaxy slowly being sucked towards the Black Hole. In fact, the closest Stars to a Black Hole should be moving towards the Black Hole very rapidly. Recent pics show Stars actually orbiting a Black Hole. If Light can't escape, how can a Star? Infinite Gravity should pull in every Star in the Galaxy. This just doesn't seem right.

So my number one complaint about Black Holes is that we can't see them. Sounds funny, but true. Rather than explain why we can't, let's look at why we should be able to. This is simple, I promise. Physicists tell us that light cannot escape a Black Hole. But Physicists also tell us that Gravity cannot bend Light. The latter being Einstein's Theory that Gravity distorts Space and it's Space that bends Light. And Physicists claim to have proved this. Now, if Gravity cannot bend Light, how can a Black Hole stop Light from escaping? We should be able to see a Black Hole! So, if one is true then both should be true, Or if one is false then both should be false. And if Light can't escape with a Photon, how can Radiation with a mass Particle? This is why I question Black Holes. Laws should be true Laws.

WARP DRIVES

For some reason this is one of the most popular questions I get asked, "Can Warp Drive exist?" Must be because I am an Engineer and if it is Man-made and it moves, I'm on it! So, you know the drill; let's look at the Mechanics and see how it is supposed to work. In Space, the Warp Drive expands Space behind the craft and compresses Space in front by distorting Space. Now this is just one version, they all involve, distorting Space or manipulating Space with large magnetic fields or matter-anti matter forces. My favorite is one that creates a wave behind the craft and the craft sort of surfs the wave following the craft. All Warp Drives require large amounts of energy to deform Space. That is the first hurdle to overcome, but not the biggest. Now I have been complaining thru two Books how Einstein's put forward the idea that Gravity distorts Space. And of course Physicists have fallen over themselves to prove Einstein right. But here's the thing; not Einstein, not a Physicist, not the Warp drive Scientists, or no one at all has defined how Space can be distorted if it is empty! Well, the short answer is that it can't. therefore if Warp drives are to work, someone will have to clearly define what Space is made of that can be distorted. Something that can be pushed against by Gravity or Electromagnetism. In short, you need to know what you are accelerating thru to design a proper engine or propulsion system. So far, nothing from anybody. Now, A few brave people have proposed the solution, an Ether or a matrix of particles in Space, but Physics is Silent. Just agreeing that Einstein's Gravity does distort Space and Warp Drives could work. So with a little help we all could working on Warp drives. And further, if I had a choice between a billion dollars a year for the Cern Collider and developing a Warp Drive, no doubt a Warp Drive would be way more important. So, Yes, Warp Drives can exist, but progress is very slow, Need more money! Just to illustrate what 3D Space looks like, here is a pic:

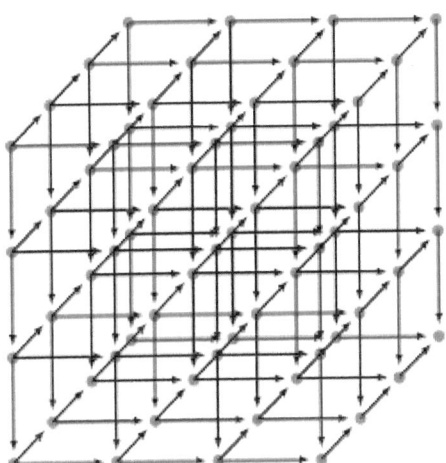

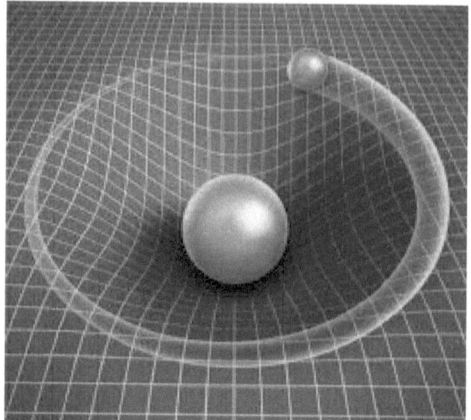

As opposed to this common pic.

Unlike the pic above this pic is 2D. Imagine a Spaceship in the cube above and try to picture how it can Warp 3D Space, especially travelling at a high rate of Speed. The Power(Energy) requirement would be Enormous, with trying to shrink Space in front and expand Space in the rear. This is one of those things I hate, a theoretical solution. If you had enough energy you could make this work, but even then, there still needs to be something in Space that can be Warped. But technology keeps improving and eventually we could make it work. But I do have to clear up a misconception: While Warp Drives could warp Space, that does not mean that Mass objects are effected. People think that if you shrink or expand Space, the Mass objects will move with it, Not So! Think of this as a Bowling Ball in a tub of water with soap bubbles on top. If you take a Vacuum and suck the bubbles toward you, the bubbles will come towards you. But the Bowling Ball will just sit there, even as the bubbles move around it. And the reverse would be true if you blew the bubbles away from you, there is no effect on the Bowling Ball. Now you could design a Spaceship to push-off of the bubbles or be attracted to bubbles, but how effective would that be? Yes, you can Warp Space, but it does not change the distance to objects. It can only be used to push off of or be attracted to. The common perception is that by compressing Space, you instantly get closer to an object. This is wrong, moving Mass objects could effect everything in the Universe, but ,as I said, Warping Space locally around Mass objects would not effect anything. Not a good concept anyways, UNLESS you can compress or expand Space Massively. This is not a next week project! And while we are on this subject, I will comment in the next Chapter on the subject that is often associated with this:

TIME TRAVEL

I am amazed at how many questions I get asked about Time Traveling and is it possible. I am an Engineer, not a Wizard or a Magician. So let's bring some discipline to this story; Life is a journey and we are all traveling through Time. Time Travel has fallen into two distinct categories:

1. The extension of life expectancy.

2. Time Jumping to completely different Periods of Time.

Life Expectancy – I don't know how this came to be confused with Time Travel, but I guess that if you live longer, you are in a Time Period that you don't belong, sort of. I think this started with Einstein's Time Dilation(Twin Paradox) and was added to with Particle Physics measuring the half-life of High Energy Particle Decay Vs. normal Particle Decay(High Energy lives longer), and finally simply Freezing people for long periods of Time. Extending one's Life expectancy is a Poor Man's Time Travel at best, since you can only go forward, but I can see the logic.

Time Jumping – This is the traditional Time Travel Theories, where people can Jump forward or back in Time at will. I have studied this subject every way that I can, but I keep running into technical difficulties. Namely, let's just say that everything in the Universe is made of atoms. All the atoms are in an exact spot at any given moment now, in the Future, and in the Past.

This becomes a real problem for going forward or rearward in time. Imagine having to have a duplicate of everything in the Universe for every moment of Time, just in your lifetime. Then, imagine if you wanted to go forward or rearward past or before your lifetime. Can you imagine all the data that would have to be stored. To put this in Science Fiction terms, every instance of Time would have to be a bubble in the core memory of some Giant Computer, that can be accessed at any time. This is a neat concept this Time Travel thing, but either every moment in Time is stored somewhere OR every moment in Time exists simultaneously, in which case it still must be stored somehow OR the easiest solution; Life is just one big Computer Program where everything is thought out and programmed. This is the easiest because there is no randomness or coincidence to deal with AND rewinding or forwarding would be just jumping to a new point in the Program. The first two examples would fail because randomness and coincidence would make the complexity difficult to overcome. The last has a chance, with some higher intelligence doing the programming, but as the population becomes larger, the Program becomes ever increasing in complexity. So highly unlikely and for technical reasons alone, I don't see this happening. And what about Time Dilation? Sort of the same thing. If you were dying of cancer, you could fly near the Speed of Light to stay alive until a cure was found. All kinds of Paradoxes would occur. Yes, that's true, and that's also why it is not happening as well. I really,

really would like to come up with a way to make Time Travel work. But the details are so complex, I'll Pass. Forget the Theories, if you go back you may get beheaded, and if you go forward you will just look stupid, get back to Reality. You have one Shot. Use it or lose it! And I was hoping this Time Travel response would not lead me to Time Dilation, but I can't see any way around it since many People think it suggests the possibility of forward Time Travel. So okay next chapter.

TIME DILATION

My least favorite chapter. This is so simple, but so misunderstood. From an Engineering standpoint, this is sideways thinking. This only exists because the math supports it. But the Physical world does not. I'll try not to go off on a rant. Why? It doesn't seem logical. If you synchronized your watch with your clock at home and drove on the expressway across Town to go shopping, then returned home and saw that your watch had lost a few minutes, What would be your first thought? Probably, maybe the battery is going dead or maybe I forgot to wind it or even, I may have to take it in and have somebody look at it. But never, Oh that Time Dilation threw my watch off again! So why, when they run tests, do they not test the accuracy of the watch(Atomic Clock)? As an Engineer, the Clock going on the Trip, should be tested first, in a Lab to simulate the actual in flight conditions. Atomic Clocks are very accurate, but they are very sensitive. Our Master Clock on Earth is kept in a vibration insulated, climate controlled room. I have never read a test for Time Dilation where they tested the Clock first against in Flight conditions. And since the deviation against the control clock is so small, this should be a standard practice. Maybe temperature, vibration, air pressure, humidity and the speed itself is causing the result. In other words; If the Clock is properly insulated for outside factors, you will not see Time

Dilation. The Earth is moving through Space at a very high rate of speed, but our Master Clock does not show any Time Dilation. No, Time Dilation is just an untested concept, because if we could see Time Dilation at 500 Miles an Hour, imagine what it would be a half the Speed of Light(186,000 Miles per Second). And why isn't Light itself affected by Time Dilation. If it is a constant, why can't everything else be?

SPACETIME

I have read about this subject over and over again, but I just can't seem to get the concept. I understand what it says is happening, but the fact is there are some things that can't be combined, because they have no effect on one another. As an example, If you stand still in Space, Time can keep on moving. And if you go the Speed of Light in Space, Time will still move at the same rate. Why? Because they are measured differently. Unfortunately for Einstein, Time does not speed up or slow down with Speed or Distance. There never has been a case where the master clock has sped up or slowed down due to movement in Space. So let me put this in a simple form that even a Physicist can understand. A distance of a 100 Light Years in Space will be 100 Light Years know matter how fast you transverse it. Conversely 10 seconds will be 10 seconds no matter what distance you cover. They are not related; they are fixed measurements set right here on Earth. In fact at slower speeds the relationship could easily be thought of as non-existant, just like it is at high speeds. So, I can't make any sense of this concept, unless it has something to do with Time Dilation. But I really don't want to go there, because it is as wrong as this is. Okay, I hoped I could answer questions about this, but there are things that are over my head. I will tell you one thing that generally hold true of any Einstein Theory. He subtlety changes reference frames to support his Theories. If you strictly stick to the concept of Earth Based Time and Distance, this Theory doesn't make sense. But if you switch to a Reference Frame where Time can slow down with Speed, it may make sense. But I don't care how fast you go, Earth's clock never slows down. Maybe next Book.

THE UNIVERSE

I get asked this question a lot. Is Space infinite? I can't say for sure, but I do have an opinion. Infinite is not a number, just a concept. That is why you will hear different word-plays on infinite, such as infinite plus 1 or an infinite set of odd numbers, etc. So, I don't believe Space or anything else is infinite. That would be too unorganized, too chaotic, too random. Whomever or whatever created this, would not have created a mess. So there is an end, somewhere, some time. But I will say this; it is big enough for me and there is more to do than can ever be done!

Ok, you ask a question, I will give you the Engineering assessment. From an Engineering view, it appears as though the Universe will continue expanding. And like many people who predict Entropy, with all the heat leaving the Universe and particles just spreading out forever, I gave that some thought. So two reasons why that won't happen;

1. The CMB shows the Universe retaining heat, albeit not very much. But as long as the fundamental Forces can still function, gravity has a chance of keeping a Galaxy together. Will all the Suns burn out? Yes, but mass does not disappear, Gravity has a chance to keep it together.

2. An advanced civilization would probably have the technology to follow the heat, meaning wherever a Star is burning, we just fly there. And maybe just move the whole planet. But just as

likely, we would learn to make our own Star. Just as people today can build a campfire, scientists will develop way to ignite small Stars or maybe even large Planets. Remember, we only need to keep our Planet warm. Now orbiting this new Star maybe a little tricky, but if we can move a whole Planet, flying into the right position won't be too hard.

So, there you have it. Our advanced Technology will save us, even if we have to fly to new parts of Space. I wouldn't worry about it too much, I see another Trillion years or at least till this simulation end!

EXPANSION

No one actually asked a question about this, but as I was writing about the Universe I had a thought about Expansion. This had it's start right after the Inflation period. It would make sense that there was a Big Bang or in my World a Huge Explosion. I say that because there is no evidence that the whole Universe was the size of a pea and none of the Laws of Physics existed; that is a Big Bang Theory. Yes, there is the CMB, but that only indicates there was an explosion, not the size of the explosion. It could have been any size explosion to throw all matter spherically into space. A huge explosion to be sure, but think of a Star going Supernovae. Shrinking Matter down to the size of a Marble or a Pea or an Atom is simply not necessary. A Galaxy sized explosion would have worked equally well. And the beauty is that all the Laws and Forces are still intact. Rewinding Expansion in reverse would only have to go as far as the time frame when Inflation Ends. When the Universe shrinks to that size, an explosion would fuel the Expansion. From that point on, the Universe develops just as it had. No Big Bang, just a convergence Matter until it compresses enough to explode. Ok, but what does this have to do with Expansion? Well, first of all, this Theory more easily explains why the Universe is not Homogenous. In the Big Bang Theories floating around, all matter would have to be created from a Super Hot Plasma, eventually cooling off and starting to form particles and Atoms.

This created Universe would tend to be very Homogenous, making it less likely to form individual Galaxies, especially of different sizes. Now, a large Mass that goes Supernova, would tend to break into pieces of all sizes, some as small as Atoms, some as large as Planets. This would easily explain why we have individual Galaxies throughout the Universe. And, Oh Yes, it can explain Expansion, too. All Mass has Gravity, especially this Mass, because it had Gravity before the Explosion; no waiting for it to form. Why is that important? Because this massive sphere was all the Mass in the Universe. The Gravity would have been intense. The massive sphere explodes and Mass objects are thrown everywhere. But they all have Gravity. And even though the explosive Force is enough to overcome Gravity, it is a drag on the spherically expanding objects. Even though many objects are pulled together by local Gravity. The Expanding sphere or Universe is all affected by the total Gravity creating a slight drag on the expansion. This gravitational effect continues until the Galaxies become further apart and the Gravity of the whole is very weak. The Galaxies, free from the drag, begin speeding up slightly. So this is Expansion, Galaxies so far apart that no outside Gravity has any effect. The inertia of the Galaxy causes it to speed up. That is my Expansion Theory and I am sticking to it.

CRACKS IN PHYSICS

Along with my views being called right out of the 1800's, there have been several comments about how Physics had to change because the Theories could not explain everything that was apparently happening in the Universe. Everything like Newton's Gravity, Einstein's failure to Unify the Universe, the Ether, The Big Bang, what is Light, Expansion and more like Dark Energy and Dark Matter and Black Holes. So here we are 100 years later and we are no closer to the answers than we were with conventional Physics. Think of That; 100 YEARS! Of Quantum Mechanics and Particle Physics, not to mention 50 Big Bang Theories and Black Hole Theories. And Expansion is not a new concept. Let's be honest before I ramble on about how many unanswered questions there are today as compared to 100 years ago. Yes, there were unanswered questions, but throughout Time there has been unanswered questions and Yes, even today there are enough to go around. So I object to the term "Cracks" or "Cracks in the Damn", when describing late 800's Physics. Why no articles on "Cracks", and because Modern Physics can't describe Dark Energy or Dark Matter and even my personal favorite "How does Gravity distort SpaceTime? It just magically does. And this distortion bends Light? What is this distortion made of?, A Brick Wall? No, it's just empty Space. Really? Nothing can be distorted! Right!

I am not criticizing the Idea that someone may have a better way
or have a new Idea, but before we throw out old Ideas, maybe we
might want to see if the new Idea actually works. Isn't 100 years
long enough to say "No, I think we'll go back to what we had
and try some new Ideas." Again, one of my favorites is this;

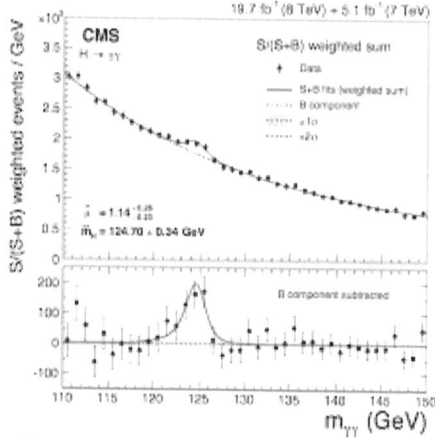

This is a Higgs Boson. Really? And this is better than an
Electron, a Proton, or a Neutron? And Yes, my Pigs can fly!
100 YEARS, and this is it? Maybe Yesterdays' Ideas with
Todays' Technology might yield some new insight.

So, Yes this is just a rant, But there must be guiding principles
like the Scientific Method and Occam's Razor. If you believe it,
Prove it! Or Shut Up. 100 Years IS ENOUGH.

Ok, last crack; why doesn't the CMB follow the shape of the
Universe? This is the CMB,

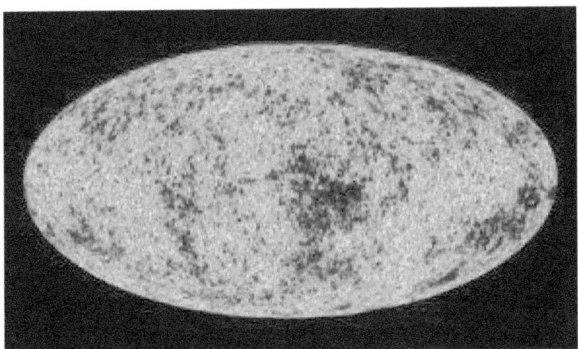

This the Universe,

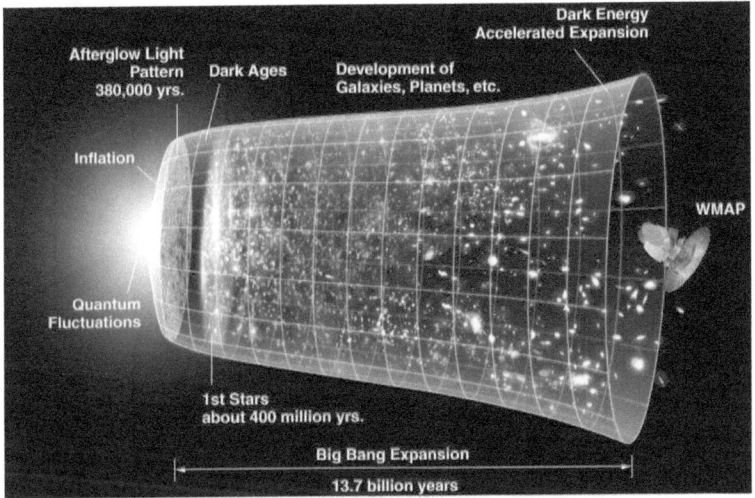

The CMB should be expanding and just like the Universe. If the Big Bang were Spherical, the CMB should be Spherical; Except that the Expansion is not Spherical, why doesn't the CMB reflect a conical Universe? OR let me put this another way; Both statements cannot be true. If the Big Bang were Spherical, as suggested by the CMB, the Expansion should be Spherical as well. But if the Big Bang was directional, as the Expansion suggests and truly is, then something is wrong in one of the Theories: A Spherical Big Bang, i.e. a marble, a pea, an atom, a singularity; A directional Expanding Universe; or The CMB data is misinterpreted. Cracks!

THE END OF PHYSICS

This Chapter is really The End of Classical Physics and the questions I get: With improved technology, why do we have as many questions today, and possibly more, as we had 100 years ago? It's very hard to clearly define When or Why things went wrong, but the general answer would be; Because Physics switched from a Linear Universe to a Digital Universe. I hate to point to a specific person, but Max Planck almost single-handedly killed Classical Physics. His solution to the Ultraviolet Catastrophe help pave the way for Quantum Physics and much more. Energy could only come in specific packets or Quanta. Very small packets to be sure, but none the less, but clearly defined discrete Packets. Now this made sense in some respects; As to why would you calculate to more than 100 decimal places for PI when 100 is sufficient for all uses? But this subtle difference, shifts us from a Linear system to a Digital System. So Who determines the size of the Packets and Why? And how do we arrive at real answers when our calculations are only approximation numbers. Yes, not important for everyday things, but when using the Size of the Universe or the size of an Electron to calculate, little differences mean a lot. So let me try to put this in perspective: If you had 10 problems you were trying to solve and they all had irrational answers, You could calculate for 100 years and never solve one. But if you could just use rational numbers, you could solve easily for all 10

Problems. Why do We think that Physics switched to Quantum Mechanics so quickly; It provides easy answers. This may not have been the intention, but it is the result. And why couldn't Classical Physics provide clear answers, this seems like a simple solution, but it applies to Theories, Testing, Experimenting, and Math and Computing: If the Assumptions are wrong, The results will be Wrong. Let's take PI as an example. The formula is; the Circumference / the Diameter = PI. But maybe the formula should be; the circumference minus .001 / the Diameter = PI. Now, this is a made up solution, but it illustrates how the Assumption could be wrong. And as a general Guideline, keep in mind that the Universe has no idea how to handle an irrational number either. So if you see an irrational number, including infinity, that should tip you off that something is wrong! Ok, back to the Death of Classical Physics. Suffice it to say, there is only one correct answer, so why not pursue it. The rule should not be to change something to make a Theory True until absolutely necessary, but to discover things to formulate the correct Theory. Quantum locks things into neat and tidy packages, but maybe Space is more quixotic. Sure, Electrons could be in neat orbitals around a Nucleus, but that implies that they all have the same energy per orbital. But in fact they could just move further away from the Nucleus linearly forming more of a cloud than orbitals. Not supporting any Theory, just pointing out that Electrons could have all amounts of energy and would logically more further away from the Nucleus Linearly, not in neat orbitals. But what is so interesting about this subject is that you would think that if one method worked so well it would be used everywhere. But that is not the case. Mega-Space is still pretty Classic and Micro-Space is mostly Quantum. Doesn't that indicate something is wrong. I don't really know why Physicists can't Unify the Universe, but it seems to me that everyone on the same page doing the same thing would be a great start. What if we could use today's technology and redo things from the 1900's, attempting to use classical Physics. Let's unify the Math.

A COLLIDER

Colliders remind me of the story of a little boy that heard that sand was one of the things it took to make glass. So he set out to find the sand in glass. He found a pane of glass in the garage and hit it with a hammer. He had hundreds of pieces of glass, but no sand. He talked to his friend and his friend said what you need is a bigger hammer, see if your dad has a sledge hammer. The boy finds the sledge hammer and proceeds to smash the glass to pieces. Again, smaller pieces of glass, but no sand. People make suggestions to help, but nothing seems to work. He has a Blacktop Roller run over it, he uses hydraulic presses, even a boulder off of a cliff, but nothing, just smaller and smaller pieces of glass. He is about to give up, but then he reads about a machine called a Collider. This sounds like just what he needs. The most powerful smasher in the World! So he gets a job as a part-timer Collider Operator. Unfortunately, the Engineers tell him that a Collider can't smash a piece of Glass, but he could smash a particle. "And you are going to enjoy this," said the engineer, "because you get to name the pieces afterward." "But do they actually MAKE things?" asked the young man. "Well no, I have been here twenty years and haven't seen anything yet. But they do have some really cool names for the pieces" Replied the Engineer. "I think I am going to like this place," said the boy, "let's smash something!" So the boy spends the rest of his life smashing particles.

Now, I am not saying that this has no value, but it would seem that this is too complex for the Universe. Smashing particles together at 99.999% the speed of Light seems like the long way around for the Universe. Imagine trying to create a whole Universe this way. And in the end you are not left with particles we have seen before, just pieces that we haven't. Occam's Razor, the simplest answer tends to be correct answer. OR An Engineering View; The more moving parts, the greater chance of something being(going) wrong. The Universe may be a much simpler place than we are led to believe. Could this generated complexity be a Crack in Quantum Physics?

CAN WE GO FASTER THAN LIGHT

I really don't like to talk this about this issue, because Physicists go berserk about this one, but since you asked, and I am an Engineer, I'll answer: Yes! This all started with Einstein, a good German Physicist. He wrote General Relativity a long while ago and said speed is relative. If you were moving one direction 60 mph and I was going the opposite direction 60 mph that would be the same thing as if one of us were stationary and the other was going 120 mph, It's all relative. So Astronomers observe Galaxies moving faster than Light, but Physicists refuse to believe it, because it is a relative speed as we are moving away at a speed and the Galaxy is expanding at a speed and that total is faster than Light, but neither is moving faster than Light. But Einstein said it is the same thing, so is Relativity wrong? Since I am an Engineer, I have no Theory to defend; So Yes we can go faster than Light. Simple logic, if two objects can move away from each other faster than Light, why can't one object go the same speed? There is no reason, except for one, and that would be a thrust engine capable of propelling you that fast. But there is nothing in Space to prevent it. But here is where Physicists get trapped by their own Theories. The general Law is: Nothing can go faster than Light. But as you investigate, you will find that this was all about Light and was just assumed for Mass objects. Or worse yet, some calculations were presented to

PROVE that Mass objects would get more massive as they approached the Speed of Light. And infinite Mass would take infinite energy to accelerate. Now, I warned you about Math, garbage in, garbage out. Of course, no one ever tested the assumption that Mass objects get more massive as they approached the Speed of Light. The Math is perfect IF the assumption is correct. Yes, Physicists, this is why Testing is required; To prove the Assumption, not the Math. The term, Garbage in, Garbage out was invented for computing, not Physical Testing. Math gives results based on inputs and manipulations of them. Computers can guarantee the accuracy of the Math, but even then, they cannot guarantee the accuracy of the assumptions. Even with 2+2=4 someone has to verify that there are, in fact, two items and two items, for the result to be valid. So don't say nothing can go faster than the speed of Light if it has never been Tested. And why is this important? This is exactly why it impossible to Unify the Universe, because every Theory has calculations. Everyone thinks they are Right! And my Pigs can Fly, too. I have calculations.

Sorry, I got a little bit off subject. So I will just tell you how we can go faster than the Speed of Light. This does just boil down to Einstein's Theory that nothing can go faster than Light. This is WRONG, WRONG, WRONG. The proper statement should be "No Electromagnetic Wave can travel faster than Light." I have explained this before, but the brief version is: Electromagnetic waves, just like electricity, need a conducter to carry them. In Space, that is The Ether. This is also what distorts by Eintein's Gravity, but unfortunately, Physicists have been unable to grasp this concept. Then why am I so sure? The evidence points to that! The general Theory reads like this "All Electromagnetic Waves are Limited to the Speed of Light, but Mass objects are Not" What evidence is there to support this Theory?:

1. It is pretty well accepted that all Electromagnetic Waves travel at the Speed of Light, from Radio Waves to Gamma Waves. All different levels of Energy, but only one speed. WHY? Could it be that the Universe itself or something in the Universe causes this.

2. But we don't see Mass objects going faster than Light. Until recently that was true. But recently Astronomers have observed Galaxies(Mass Objects) moving away at speeds faster than Light. Regardless of the propulsion system, this is true.

3. If Expansion is causing faster than Light Speeds for Mass Objects, according to Physicists, we should see the increase in Light as well as Mass, but we don't. Why are the Speeds different? Light Speed could be restricted by an Ether, but Mass Speed are additive, No Question. A light shining from a moving train will only go the Speed of Light, No Question. A ball thrown from a moving train will travel the speed of the train PLUS the speed of the throw, No Question. Something must be causing this affect, No Question.

THEREFORE: Electromagnetic Waves are limited by an Ether. Mass objects are unaffected by the Ether.

The conclusion is: Yes we can go faster than Light. We just need a faster than Light propulsion System. Well I can hear the Physicists already, there is no propulsion system faster than Light. Oh, but there is Grasshoppers. Remember the example of a train's speed and a thrown ball speed being additive? Well what if the train were going the Speed of Light and then you threw the ball. What would be the Speed of the ball then? This is just one of a million ways we could generate. faster than Light Speed. Strangely enough, there is a working example of this method. To Speed power a SpaceShip sufficiently to travel the Galaxy, it has been proposed to use Nuclear Bombs to increase the speed. Each Exploding Bomb would cause the ship's speed to increase. This is a perfect example of speed being additive for Mass Objects. Imagine if they took this Method all the way to the Speed of Light, what would happen?

E=MC² ?

I actually know how this question came about, but I have very little data to answer it. How it came about is that in a Hydrogen Bomb, only about .7% of the mass is transferred to energy. Now that doesn't sound like a big deal, but the question is; What happened to the Rest of the Bomb? The Bomb gets blown apart in the explosion so we have little way of knowing, but we can estimate the diameter of the explosion by the extent of the destruction. We can then calculate what sized the explosion could have been by the amount of material, and compare the two, actual vs calculated. And gives the result that only .7% of the material produced energy. Again; What went wrong?

Here is what we know: 100% of the fissionable material was prepared in the same way and all could split and give off energy. The triggering device, although not perfect, provided enough Neutrons to split 100% of the material.

So there are only 3 conclusions that we can come to:

1. The material was never 100% fissionable.

2. The trigger was never intended to be 100% successful.

And since the first two to would have risk, unacceptable for a Nuclear Bomb. Redundancy should be built in. The possible solution is,

3. $E \neq MC^2$. I don/t have enough data to support any point of view, but I can comment. It has always bothered me that C is the Speed of Light. Why is the conversion Constant of matter to

energy exactly the same as Light Speed. They are two different things. Even if you believe they are both Constants, why would they be, or need to be, the same. Yes, both the same would tidy up some formulas, but that is not Proof, just a Wish. We wouldn't use absolute Zero as the Constant, why the Speed of Light? In fact, with the amount of Energy given of by such a large amount of matter in the Atomic Bombs, C seems like it should be smaller. Ok, with such little data, all I can say is that it just doesn't seem right. Why, because what is missing is the energy that we must add to get energy out. It goes like this; An atom of U-235 gets a Neutron added to it, which causes it to split and give off more Neutrons, which hits more U-235 atoms and so on, a chain reaction. And, of course, every time an atom splits it gives of energy. Sounds simple but something is wrong, the numbers don't seem to add up. The first Atomic Bomb was 64 Kg of U-235 and only less than 1 Kg actually fissioned. If you believe the chain reaction theory, it all should have fissioned, what would have stopped the chain reaction? And why? So, there are only 2 possiblities, the chain reaction stopped, possibly on its own, OR The fission was complete but the yield was way less than $E=MC^2$! Why would they build a Bomb with a triggering mechanism that only triggered less than 1% of the U-235? The more likely scenario would be that the triggering mechanism worked perfectly, but the bomb yield was way less than predicted. Same thing with the next Bomb. Does

$E=MC^2$? Not enough data, but from an Engineering View, something is wrong.

An Atomic Bomb. If $E=MC^2$ this Bomb would only need to be $1/10^{th}$ the size, possibly $1/100^{th}$ the size.

SCIENTIFIC METHOD

I have included this in every book as I have stated that much of Physics just ignores the Scientific Method and has more of a "If I can think it, Therefore it is." Attitude. But now I have come to the realization that of course they don't follow the Scientific Method, They are "Theoretical" Physicists. What do I mean by that? For example, Einstein stated that Time Dilation existed, but did he ever actually observe Time Dilation or test Time Dilation BEFORE he published it. And in his Theory, he never gave the mechanics of how or why Atoms would speed up or slow down based on speed. Same thing with Einstein's Gravity; did Einstein actually observe Space being warped, did he test for it? And what are the mechanics; What in empty Space can be warped by Gravity? Now before 100 Physicists jump up and say these Theories have been tested over and over and proved right, Yes, but not before Einstein published the Theories and not for many years after and not by Einstein himself. Scientists are supposed to follow the Scientific Method as they are discovering and proving the Theory, not several years later. I have no problem with "Thought Experiments" and thinking "What if", But when you publish it as reality I draw the line. No, No, and No! I know Pigs can Fly, really. Publish it as Science Fiction if you like, but don't publish it as a Scientific Paper as though it is reality. How many Big Bang Theories are out there? Time Travel, Worm Holes, and Warp Drives. Sub-Atomic Particles, Black Holes, Photons; Show me the Proof or Shut Up! Follow the Scientific Method.

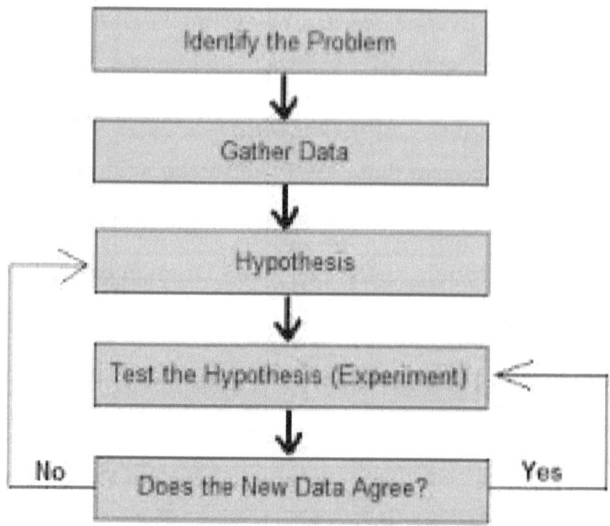

Don't show me all your Math, show me some test data. Why? Because even in the 1600's when the first Scientific Method was developed, they knew Math was just a language and like any language, you can tell any story you want. Or as my Computer Programming Teacher used to say, "Garbage in, Garbage out!"

APPENDICES

This is my favorite Pic:

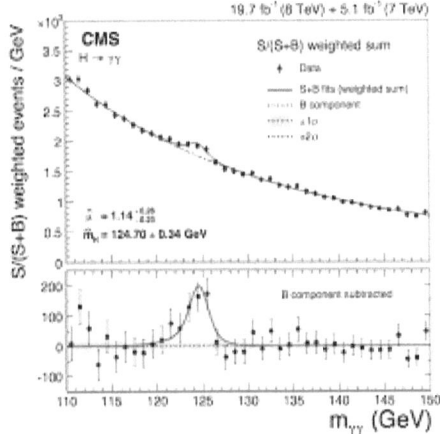

It won a Nobel Prize. This is what SHOULD have won a prize:

NOTES

This is it for questions, so far. My Universe is very simple and easy to understand. My guiding light through out has been, of course, Occam's Razor, The Simplest Solution is Usually the Correct Solution. Why not for the entire Universe. Oh yes, God had a hand in it too. There are some subjects that are just way over my head. My best thought; Mass objects can go faster than Light. Why, because for Mass objects Speed is additive; Proved many times over. So we just need to chain propulsion systems that add propulsion to over Light Speed. And for you Sci-Fi fans; Yes you could even use some type of slingshot effect. Although my next book will probably be a children's Christmas Book, I will keep an eye on the Universe in case anything interesting comes along, like going to Mars.

ACKNOLEDGEMENT

A special thanks to Keisan Casio Triangle and Advanced online calculators. Without them I could not have found a solution for PI. The triangle calculators are good to 50 digits and the advanced calculator is good to 102. This allowed calculations to a Sextillion in seconds. Great Product. And the support is top notch. No this was not just what every one else is doing, dividing the circumference by the diameter, this was a quest to find what was wrong. These calculators made is easy to do 100's of calculations in minutes.

www.ingramcontent.com/pod-product-compliance
Lightning Source LLC
Chambersburg PA
CBHW021003180526
45163CB00005B/1879